discovermore
Earth's Processes

What Is the Rock Cycle?

Doug Bradley

Britannica Educational Publishing
IN ASSOCIATION WITH
THE ROSEN PUBLISHING GROUP

Published in 2027 by Britannica Educational Publishing (a trademark of Encyclopædia Britannica, Inc.) in association with The Rosen Publishing Group, Inc.
2544 Clinton Street, Buffalo, NY 14224

Distributed exclusively by Rosen Publishing.
To see additional Britannica Educational Publishing titles, go to rosenpublishing.com.

Editor: Greg Roza
Book Design: Deanna Lepovich

Photo Credits: Cover JStagg_Photography/Shutterstock.com; (series background) Dai Yim/Shutterstock.com; p. 4 Tetra Images, LLC/Alamy Photos; p. 5 Roberto Nistri/Alamy Photos; p. 7 (top) Tartila/Shutterstock.com; p. 7 (bottom) Endless Travel/Alamy Photos; p. 8 Susan E. Degginger/Alamy Photos; p. 9 José María Barres Manuel/Alamy Photos; p. 11 (top) VectorMine/Shutterstock.com; p. 11 (bottom, inset) Dorling Kindersley ltd/Alamy Photos; p. 11 (bottom) Photology1971/Alamy Photos; p. 12 Glyn Thomas Photography/Alamy Photos; p. 13 John Zada/Alamy Photos; p. 14 Olga Borsh/Alamy Photos; p. 15 Anna Vereshchak/Alamy Photos; p. 17 (top) Fossil & Rock Stock Photos/Almay Photos; p. 17 (bottom) Avalon.red/Alamy Photos; p. 19 (top) robertharding/Alamy Photos; p. 19 (bottom) Cynthia Farmer/Alamy Photos; p. 20 Graham Mulrooney/Alamy Photos; p. 21 JHVEPhoto/Alamy Photos; p. 22 Jaber Yazdankhah Kenary/Alamy Photos; p. 23 Nick Greaves/Alamy Photos; p. 25 (top) Lynne Evans/Alamy Photos; p. 25 (bottom) Valery Voennyy/Alamy Photos; p. 26 Luis Fonseca/Alamy Photos; p. 27 Patti McConville/Alamy Photos; p. 28 Gabbro/Alamy Photos; p. 29 Philip Dickson/Alamy Photos.

Library of Congress Cataloging-in-Publication Data

Names: Bradley, Doug, 1971- author
Title: What is the rock cycle? / Doug Bradley.
Description: [Buffalo] : Britannica Educational Publishing, [2027] |
Series: Discover more: Earth's processes | Includes index.
Identifiers: LCCN 2026004603 (print) | LCCN 2026004604 (ebook) | ISBN 9781641906203 library binding | ISBN 9781641906197 paperback | ISBN 9781641906210 ebook
Subjects: LCSH: Geochemical cycles--Juvenile literature | Rocks--Juvenile literature
Classification: LCC QE432.2 .B725 2027 (print) | LCC QE432.2 (ebook)
LC record available at https://lccn.loc.gov/2026004603
LC ebook record available at https://lccn.loc.gov/2026004604

Printed in China

Some of the images in this book illustrate individuals who are models. The depictions do not imply actual situations or events.

CPSIA Compliance Information: Batch #QSBRIT27. For further information contact Rosen Publishing at 1-800-237-9932.

Find us on

Contents

Earth's Changing Landscape 4
Rocks and Minerals 6
In the Cycle 10
Hot Rocks! 12
What Are Crystals? 16
Flat Rocks 18
Stacking Up 22
Changing Rocks 24
Geology and the Rock Cycle 28
Glossary 30
For More Information 31
Index 32

Earth's Changing Landscape

Earth's landscape has changed greatly over the course of millions of years. Over time, rocks have been formed, worn down, and formed again. This is called the rock cycle. A cycle is a series of events or actions that repeat regularly. There are several cycles in nature. The seasons are a cycle. The way water moves around Earth is called the water cycle.

Earth's rocky landscapes may seem like they never change. However, they are slowly yet constantly undergoing a cycle of change.

This is a trilobite fossil. Trilobites were sea creatures that lived long ago. This fossil is proof that the land was once underwater.

Scientists can find clues to Earth's past by studying how its rocks have changed or moved. For example, rocks can show us that large areas of land were once covered by thick layers of ice. Layers of rock can show us how ancient earthquakes shaped mountains and valleys. The tiny **fossils** of ocean creatures found in rocks can show us how some areas that are now dry land were once covered by water.

WORD WISE

FOSSILS ARE THE REMAINS OR TRACES OF PLANTS AND ANIMALS THAT LIVED LONG AGO.

Rocks and Minerals

Rock is a hard, naturally occurring material made up of one or more minerals. Minerals are inorganic substances, meaning that they do not come from an animal or a plant. Rock makes up the outer layer of Earth, called the crust. At its thickest, the crust is about 19 miles (31 kilometers) thick. The lower parts of this layer are solid rock, or bedrock. Broken bits of rock lie on top. The tiniest bits of rock make up sand and soil. Underneath Earth's crust is another layer of mostly solid rock, called the mantle. Pools of melted rock, or magma, form in the upper part of the mantle and the lower part of the crust.

Scientists divide rock into three main types. These are igneous, sedimentary, and metamorphic rock. Each of these types of rock forms in a different way and looks different from each other.

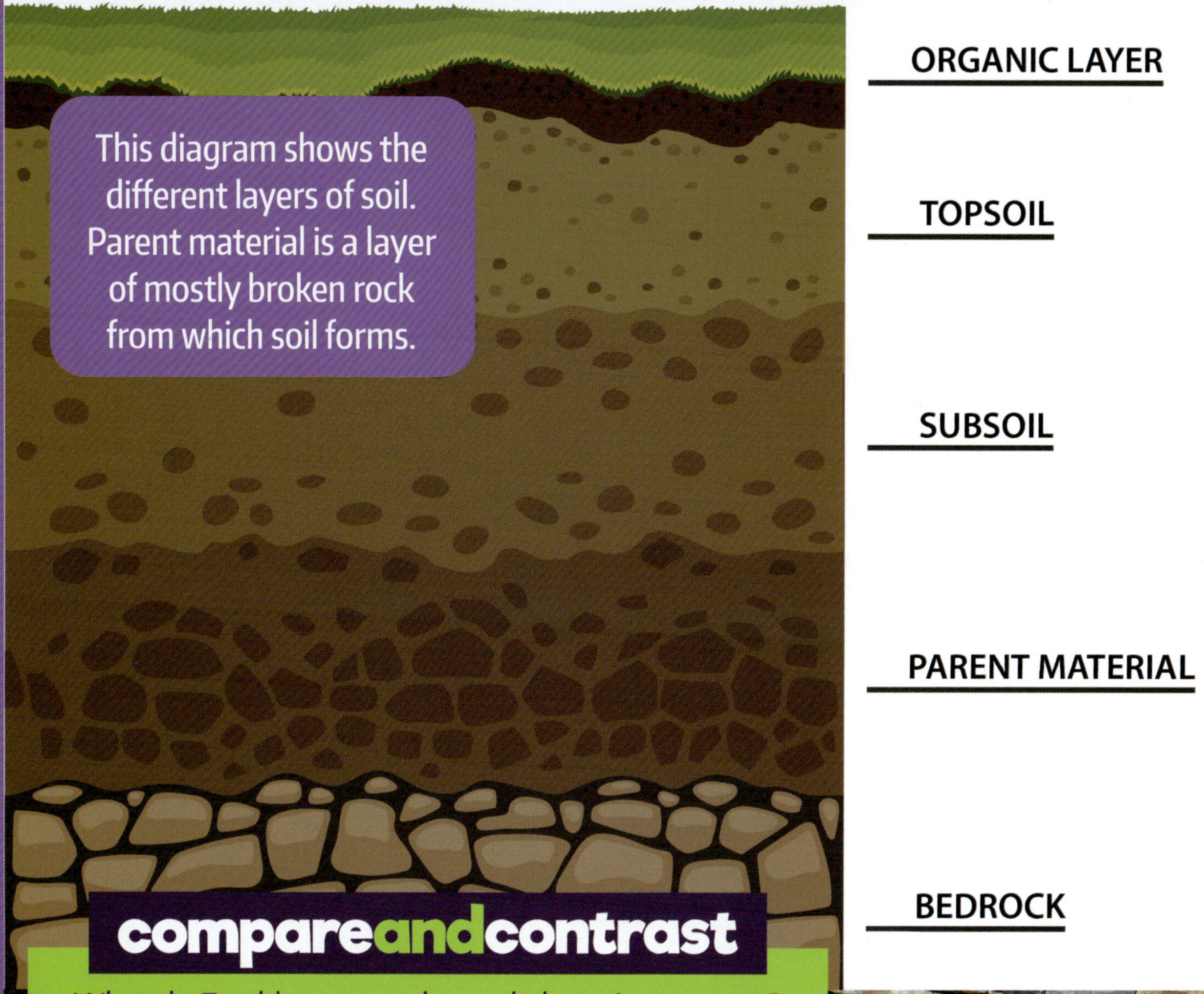

This diagram shows the different layers of soil. Parent material is a layer of mostly broken rock from which soil forms.

compareandcontrast

What do Earth's crust and mantle have in common? How are they different?

Slate is a metamorphic rock that can be cut into thin layers. It has long been used to make roof tiles.

Minerals are found on Earth's surface as well as deep underground. They make up Earth's rocks, sands, and soils. Most rocks contain several types of minerals. For example, the rock granite is largely composed of the minerals quartz, feldspar, and mica.

Mineralogists, or people who study minerals, have identified hundreds of minerals. Some of the most common minerals are metals—for example, gold, silver, copper, and platinum. Diamond, quartz, sulfur, mica, talc, and salt are other well-known minerals.

Quartz is a common mineral made of silicon and oxygen.

Shown here is a mineral streak test for the mineral hematite, also called iron oxide.

Mineralogists use certain properties to identify minerals: color, **luster**, hardness, cleavage, and streak. Cleavage is how minerals split or break. Streak is the color a mineral leaves when rubbed against rough porcelain. Mineralogists also test how minerals react to electricity and magnets.

WORD WISE

LUSTER DESCRIBES HOW A MINERAL APPEARS AS IT REFLECTS LIGHT. SOME MINERALS ARE TRANSPARENT, OR SEE-THROUGH. OTHERS ARE IRIDESCENT, WHICH MEANS THAT THEIR COLOR CHANGES AS LIGHT HITS THEM FROM DIFFERENT DIRECTIONS.

In the Cycle

The rock cycle consists of several natural processes that form, break down, change, and reform rocks over time. Temperature, pressure, and changes in conditions at and beneath Earth's surface drive the rock cycle.

The process of weathering wears rocks down, while the erosion process carries away weathered bits of rock. Over millions of years these bits of rock build up, harden, and become solid sedimentary rock. If that rock becomes deeply buried, it may melt into magma. When magma cools down, it becomes igneous rock. Heat and pressure inside Earth's crust can shape old rock into metamorphic rock.

The rock cycle takes many millions of years, and rocks can follow many paths through the cycle. Any of the three types of rock can change to form any other type of rock under the right conditions.

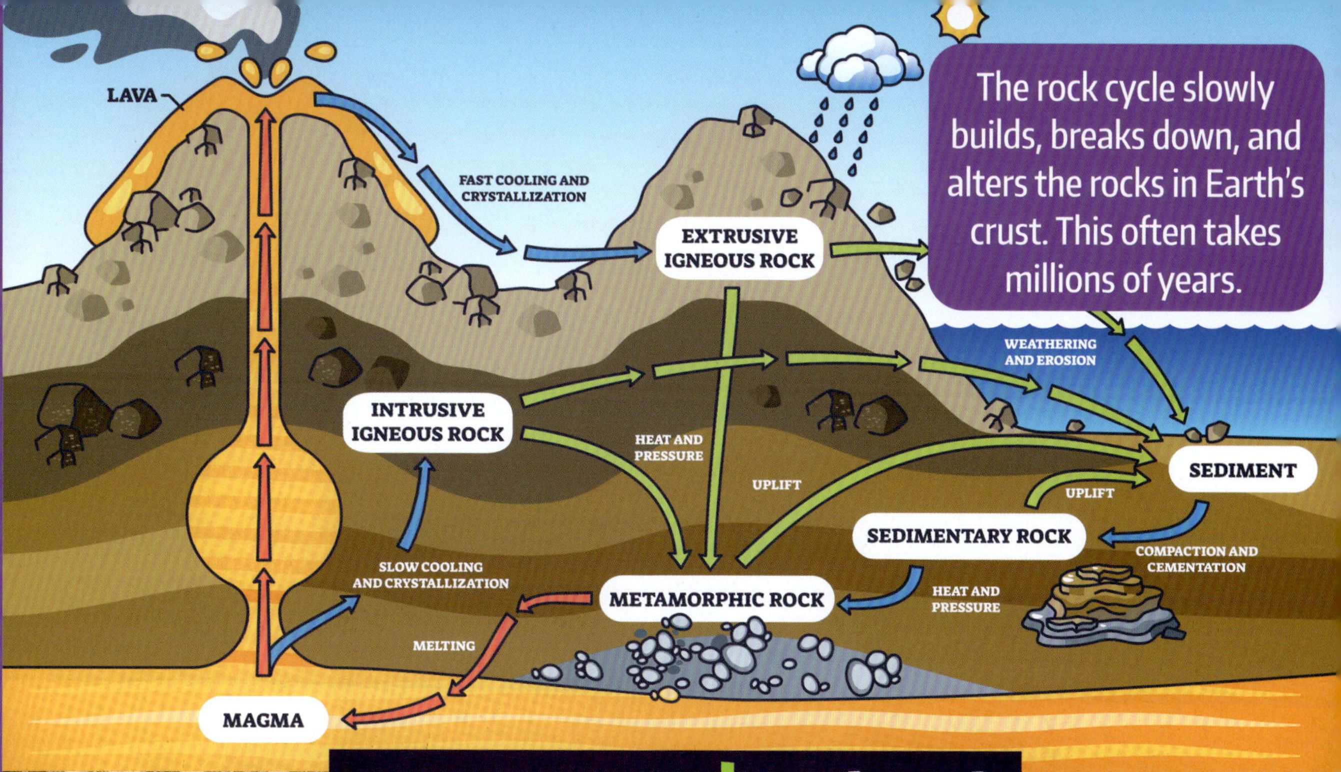

The rock cycle slowly builds, breaks down, and alters the rocks in Earth's crust. This often takes millions of years.

compareandcontrast

How is sedimentary rock similar to igneous and metamorphic rock? How is it different?

Limestone is a sedimentary rock. When pressure and heat is applied to limestone inside Earth, it transforms to the metamorphic rock marble.

Hot Rocks!

The word "igneous" comes from the Latin word ignis, which means "fire." Igneous rocks got their name because the magma from which they form is so hot. Magma is a soft, fluidlike substance. It is formed inside Earth, where the temperatures are very hot. Magma exists at temperatures of 1,100° to 2,400° F (600° to 1,300° C).

Volcanic eruptions can be very destructive. However, they also help create new igneous rock.

Obsidian is a black, glassy mineral. Ancient people used obsidian to make sharp knives and arrow points.

Magma is less dense than the surrounding solid rocks, so it rises toward the surface. It may settle within the crust or erupt at the surface from a volcano. If it erupts it becomes a **lava** flow. When magma cools down, it returns to a solid state and become igneous rock.

Most of Earth's crust is made up of igneous rock. Examples of this type of rock include obsidian, granite, and basalt.

WORD WISE

AFTER MAGMA BREAKS THROUGH EARTH'S CRUST, IT IS CALLED LAVA.

The two main types of igneous rock are intrusive and extrusive rock. Intrusive igneous rock forms when the magma cools slowly, usually deep underground. Granite is an intrusive igneous rock.

Granite looks different depending on the other minerals it contains.

Pumice is a lightweight extrusive rock that forms when lava cools quickly and traps air bubbles.

Extrusive igneous rock forms when magma cools very quickly. This tends to happen when it erupts from a volcano as lava. The air and water aboveground are much cooler than the lava. Basalt is an example of this type of rock. Some extrusive rock is formed from pieces of debris that are thrown out of a volcano during an eruption. If there are no minerals present, the lava cools quickly to form smooth, glassy rocks.

Consider This

The most common igneous rock on Earth is basalt. Why do you think that is?

What Are Crystals?

In science, a crystal is a certain type of solid object. Scientists describe matter, or everything that can be seen, according to the way it is made. All matter is made up of tiny parts called atoms and molecules. If those parts are arranged in a regular pattern, then the object is a crystal. On the outside, crystals have a regular pattern of flat surfaces that meet in sharp corners.

Most igneous rocks are made of crystals. Magma usually forms crystals when it cools. As intrusive rock cools slowly, it has time to form large crystals. This results in coarse- or medium-grained rocks.

When extrusive igneous rock is formed, the lava cools so quickly that crystals have little or no chance to grow. The rocks that form may be fine-grained. The minerals might be so small that a microscope is required to view them.

Granite is rich in quartz and feldspar. It can be a coarse- or medium-grained igneous rock.

compareandcontrast

How are intrusive and extrusive igneous rocks similar? How are they different?

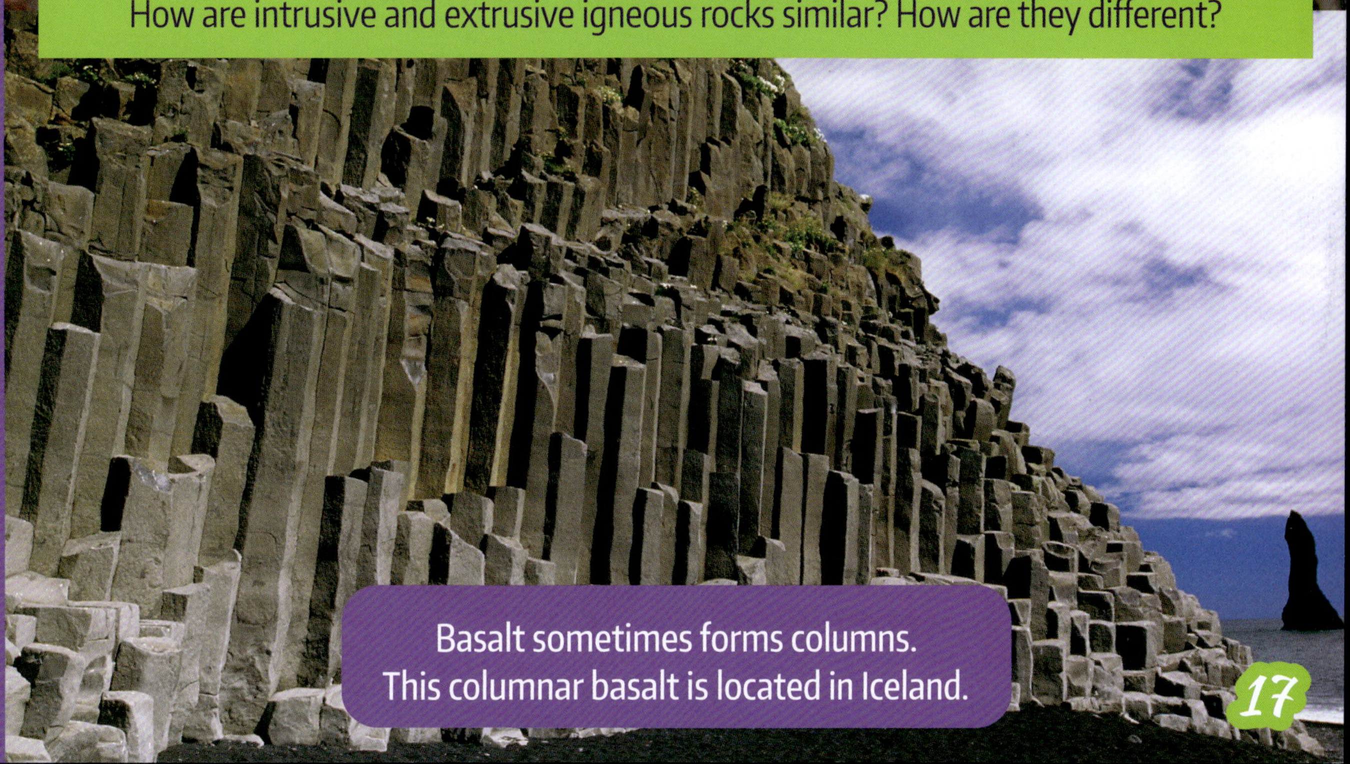

Basalt sometimes forms columns. This columnar basalt is located in Iceland.

Flat Rocks

Sedimentary rock is often found in layers or bands. They can be formed in several ways. Some is formed through erosion. This happens when rocks of any kind are worn away by rain, wind, and other elements. Tiny fragments of the original rock, along with dissolved minerals, are carried away to another site. There, they may combine with other materials, such as mud. This settled matter is called sediment.

As layers of sediment build up over thousands of years, they are compressed, or squashed, and become bonded together. Sometimes the fragments are too big to be compressed. In that case, the spaces between the fragments are filled in by the dissolved minerals, which make a kind of cement when they harden.

Sedimentary rocks can be put into groups according to the way they are created and the size of the fragments that they contain. For example, conglomerates are made of larger particles with minerals between them acting as cement to hold them together.

You can see the sedimentary layers in these sandstone formations in the Arizona desert near Glen Canyon.

compareandcontrast

Sandstone and coquina are very different looking, but they are both sedimentary rocks. How are they the same? How are they different?

Coquina is a sedimentary rock largely made up of animal shells cemented together.

Limestone is a sedimentary rock that can form in shallow water. When the surrounding air is very dry, the water may evaporate, or dry up, leaving behind minerals that harden into rock.

Travertine is a type of limestone rock that forms from mineral-rich water in hot springs.

The White Cliffs of Dover is a famous limestone formation in southern England. It is made of a light limestone called chalk.

Sedimentary rock can also form when the shells of sea creatures fall to the floor of the sea after the creatures die. When the shells form a thick enough layer, they become compressed and form rock. This is called limestone.

Limestone is used to make cement, roads, and fertilizer. Many forms of limestone are used as building materials. People use limestone for making floors, for covering the walls of buildings, and for making monuments. Limestone is useful for building because it does not easily wear away and it is not difficult to shape.

Consider This

Limestone deposits often contain fossils. How might limestone help scientists learn more about Earth's past?

Stacking Up

Sedimentary rock forms in layers. Each layer may have a different color, particle size, or type of cement. Scientists study these layers. They can figure out what kind of sediment each layer is made from, as well as how it was carried to the spot where it hardened into rock. The deepest layers are the oldest. More recent layers are closer to the surface. Each layer contains unique types of rocks and fossils. These give scientists many clues about Earth's history.

After many millions of years, Earth's forces have warped these layers of sedimentary rock into fascinating shapes.

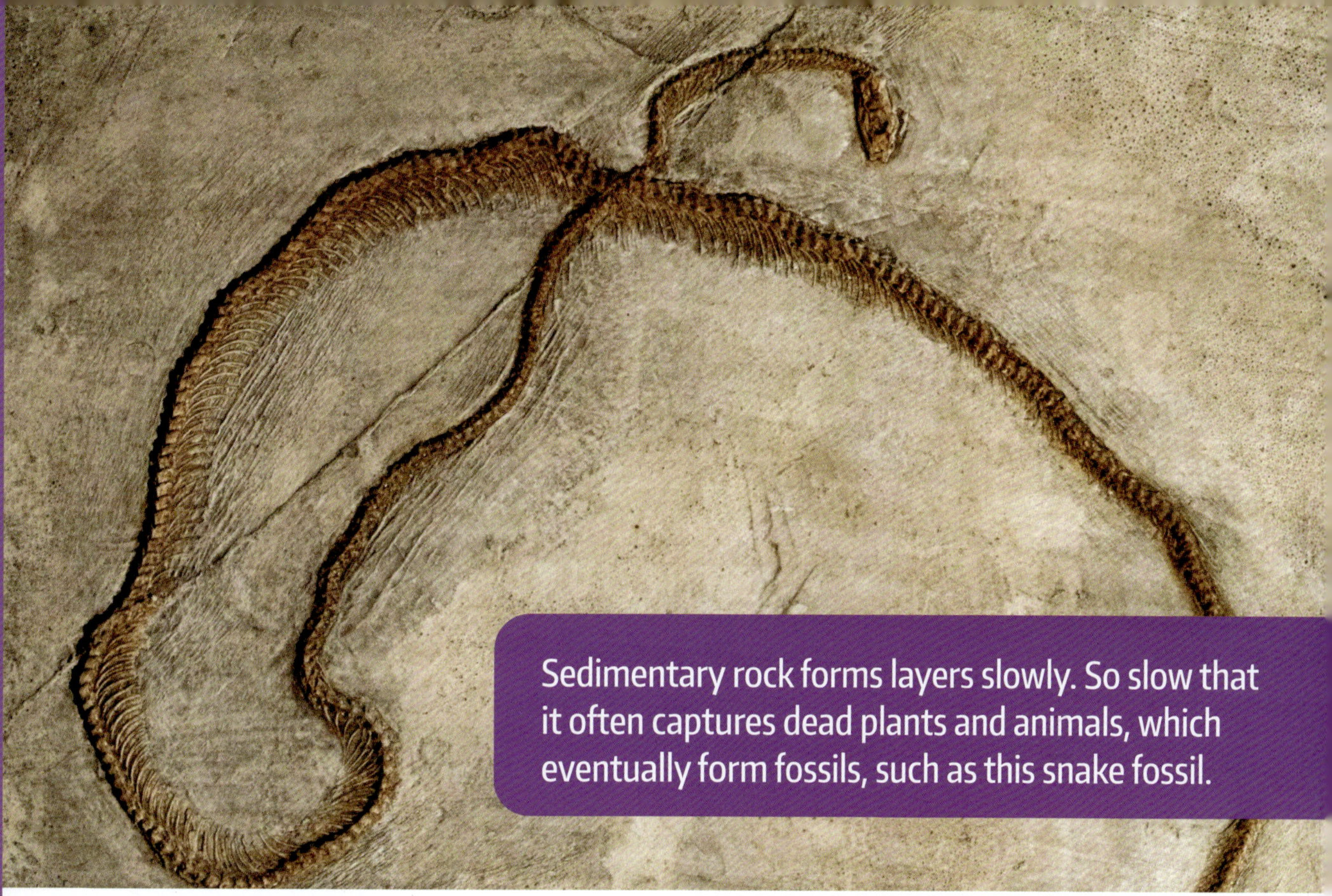

Sedimentary rock forms layers slowly. So slow that it often captures dead plants and animals, which eventually form fossils, such as this snake fossil.

Layers of sedimentary rock are also important because they contain important natural resources, such as salt deposits and groundwater. The world's store of **fossil fuels** (oil, natural gas, and coal) is also buried in sedimentary rock deep underground.

WORD WISE

FOSSIL FUELS ARE THE REMAINS OF ORGANISMS THAT LIVED LONG AGO. THEY ARE USED ACROSS THE WORLD TO MAKE ELECTRICITY, HEAT HOMES, AND POWER VEHICLES.

Changing Rocks

Metamorphic rocks form when other types of rocks undergo great heat and pressure inside Earth. These rocks do not melt, but they do form new minerals, textures, and crystal structures. Sometimes, the minerals get separated into visible bands. The name metamorphic comes from Greek words meaning "change of form."

Metamorphic rock can form from igneous rock, sedimentary rock, or another metamorphic rock. The most common metamorphic rock is perhaps gneiss (pronounced "nice"), which can form from granite under intense heat and pressure. Marble and slate are two more examples of metamorphic rocks.

Studying the chemicals that make up metamorphic rock can reveal the original rock type. Limestone and marble are both made of calcium carbonate. The sedimentary rock limestone can become marble with enough heat and pressure.

Northern Scotland is a rocky and sparsely populated area. Many landforms in the area are made largely of banded gneiss.

compareandcontrast

How are the bands in metamorphic rock like the layers in sedimentary rock? How are they different?

Hornfels is a metamorphic rock formed when magma heats up and changes other rocks around it.

Metamorphic rock requires great heat and pressure to form. It often forms in places where large sections of Earth's crust called plates come together and push against each other. When two plates collide, great temperatures and pressures can be produced. Such plate collisions can cause new mountain ranges to form. Most of the world's mountain belts are at least partly made of metamorphic rocks. Spectacular examples are provided by the Highlands of Scotland, the Alps in Europe, the Himalayas in Asia, and the northern Appalachians in North America.

Cerdeira is one of 27 "schist villages" in central Portugal. All the buildings are made of schist. Most of these villages had been "lost" until recently rediscovered and repaired.

Visitors to Central Park in Manhattan, New York, can see and even climb on schist formations. This type of rock, which is actually ancient bedrock, is known as Manhattan schist.

Metamorphic rock can also form in places where magma is forced through existing rocks, heating them and increasing the pressure. Sometimes rocks are changed when layers of overlying heavy rocks put huge pressure on them.

Consider This

Metamorphic rocks most often form deep within Earth rather than near or on the surface. Why do you think that is?

Geology and the Rock Cycle

Geology is the study and history of Earth's physical features. It's an important science for many reasons. It answers questions about how Earth came to have its present shape and form. It is useful for finding important materials in Earth's crust, such as oil. Geology is also helpful for predicting earthquakes and other natural hazards.

This group of geologists is studying rocky landforms on Iceland.

Do you think you'd like to be a geologist one day? You can get a head start by trying rock collecting. You may even find fossils!

Scientists who work in geology are called geologists. It is thanks to them that we know about the rock cycle. You don't need a degree in geology to learn more about rocks, though. Collecting rocks and minerals is a popular hobby. Rock collectors call themselves rockhounds. Rockhounds usually have a basic understanding of the rock cycle and how the rocks they find were formed. They use a number of simple tests to help identify their finds. A magnifying glass and rock hunter's guidebook are useful, too.

Consider This

Where are some good places to search for rocks?

Glossary

atom One of the tiny particles that are the building blocks of all matter.

bedrock The solid rock lying under surface materials (such as soil) that are not in layers.

cement A binding element or substance.

crust Earth's thin, rocky outer layer.

crystal A solid form of a substance or mixture that has a regularly repeating internal arrangement of its atoms and often external plane faces.

deposit A collection of mineral matter in nature.

evaporate To turn from a liquid to a gas.

fragments Parts that are broken off.

groundwater Underground water that supplies wells and springs.

mantle The layer of Earth lying between the crust and the core.

molecule The smallest unit of a substance that has all the properties of that substance. Molecules are made of atoms.

monument A building, pillar, stone, or statue honoring a person or event.

porcelain A kind of pottery.

pressure A force that can move or press down on things.

properties Special qualities or features.

temperature The degree of hotness or coldness of something (such as air, water, or the body) as shown by a thermometer.

weathering A natural process that slowly breaks apart or changes rock.

For More Information

Books

Clamant, A. B. *The Rock Cycle*. Buffalo, NY: Norwood Press, 2025.

Tomecek, Steve. *National Geographic Kids Ultimate Rockopedia*. Washington, D.C.: National Geographic Kids, 2020.

Websites

Mineral Facts for Kids
kids.kiddle.co/Mineral
Read more about minerals and see many examples of them.

Understanding Rock Collecting
fossibelle.com/articles/understanding-rock-collecting-guide/
Do you want to learn more about rock collecting? This thorough guide will teach you all you need to know.

Index

B

bedrock, 6, 27

C

collecting rocks, 29
crust, 6, 7, 10, 11, 13, 26, 28
crystals, 16, 17, 24

E

earthquakes, 5, 28
erosion, 10, 18

F

fossils, 5, 21, 22, 23
fossil fuels, 23

G

geology, 28, 29

H

heat, 10, 11, 24, 25, 26, 27

I

igneous rock, 6, 10, 11, 12, 13, 14, 15, 16, 17, 24

L

landscapes, 4, 25, 28
lava, 13, 15, 16

M

magma, 6, 10, 12, 13, 14, 15, 16, 25, 27
mantle, 6, 7
metamorphic rock, 6, 7, 10, 11, 24, 25, 26, 27
minerals, 6, 8, 9, 13, 14, 15, 16, 18, 20, 24, 29

P

plants, 5, 6, 23
pressure, 10, 11, 24, 26, 27

S

sedimentary rock, 6, 10, 11, 18, 19, 20, 21, 22, 24, 25

V

volcano, 12, 13, 15

W

weathering, 10